Weather on Earth

Developed at
The Lawrence Hall of Science,
University of California, Berkeley
Published and distributed by
Delta Education,
a member of the School Specialty Family

© 2012 by The Regents of the University of California. All rights reserved. No part of this book may be reproduced or transmitted in any form or by any means, electronic or mechanical, including photocopying or recording, or by any information storage and retrieval system, without permission in writing from the publisher.

1325254
978-1-60902-046-0
Printing 1 — 3/2012
Quad/Graphics, Versailles, KY

Table of Contents

Investigation 1: What Is Weather?
What Is Air? 3
Earth's Atmosphere 7
Weather Instruments 14

Investigation 2: Heating Earth
Uneven Heating 17
Heating the Air: Radiation and Conduction 21
Wind and Convection 27
Wind Power 32
Solar Technology 34

Investigation 3: Water Planet
Condensation 43
Where Is Earth's Water? 47
The Water Cycle 48

Investigation 4: Weather and Climate
Severe Weather 53
Weather Maps 62
Earth's Climates 71
Global Climate Change 76

References
Science Safety Rules 84
Glossary 85
Index 88

Clouds float in air.

What Is Air?

You can't see it or taste it. You can't even smell it. But you might feel it as a gentle breeze brushing across your skin. **Air** is difficult to understand because it is not easy to observe with your senses. Is air one thing or a mixture of things? And where is air? Is it everywhere or just in some places?

As we go about our everyday lives, we usually travel with our feet on solid **Earth** and our heads in the **atmosphere**. The atmosphere is all around us, pressing firmly on every part of our bodies—top, front, back, and sides. Even if we attempt to get out of the atmosphere by going inside a car or hiding in a basement, the atmosphere is there, filling every space we enter.

An atmosphere is the layer of **gases** surrounding a planet or star. All planets and stars have an atmosphere around them. The **Sun's** atmosphere is mostly hydrogen. Mars has a thin atmosphere of carbon dioxide (CO_2) with a bit of nitrogen and a trace of **water vapor**. Mercury has almost no atmosphere at all. Each planet is surrounded by its own mixture of gases.

A view from space, looking down through Earth's atmosphere

Earth's atmosphere is made up of a mixture of gases we call air. Air is mostly nitrogen (78 percent) and oxygen (21 percent), with some argon (0.93 percent), carbon dioxide (0.039 percent), **ozone**, water vapor, and other gases (less than 0.04 percent together).

Nitrogen is the most abundant gas in our atmosphere. It is a stable gas, which means it doesn't react easily with other substances. When we breathe air, the nitrogen goes into our lungs and then back out unchanged. We don't need nitrogen gas to survive.

Oxygen is the second most abundant gas. It makes up about 21 percent of air's volume, and it accounts for 23 percent of air's mass. Oxygen is a colorless, odorless, and tasteless gas. Oxygen combines with hydrogen to form water. Without oxygen, life as we know it would not exist on Earth.

Oxygen and nitrogen are called permanent gases. The amount of oxygen and nitrogen in the atmosphere stays constant. The other gases in the table are also permanent gases, but are found in much smaller quantities in the atmosphere.

| Permanent Gases of the Atmosphere ||
Gas	Percentage by volume
Nitrogen	78.08%
Oxygen	20.95%
Argon	0.93%
Neon	0.002%
Helium	0.0005%
Krypton	0.0001%
Hydrogen	0.00005%
Xenon	0.000009%

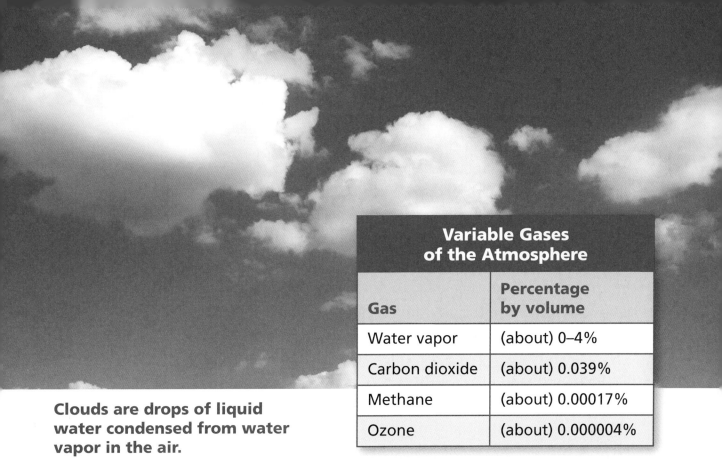

Variable Gases of the Atmosphere	
Gas	Percentage by volume
Water vapor	(about) 0–4%
Carbon dioxide	(about) 0.039%
Methane	(about) 0.00017%
Ozone	(about) 0.000004%

Clouds are drops of liquid water condensed from water vapor in the air.

Air also contains variable gases. The amount of each variable gas changes in response to activities in the environment.

Water vapor is the most abundant variable gas. It makes up about 0.25 percent of the atmosphere's mass. The amount of water vapor in the atmosphere changes constantly. Water moves between Earth's surface and the atmosphere through **evaporation**, **condensation**, and **precipitation**. You can get an idea of the changes in atmospheric water vapor by observing **clouds** and noting the stickiness you feel on your skin on humid days.

Carbon dioxide is another important variable gas. It makes up only about 0.039 percent of the atmosphere. You can't see or feel changes in the amount of carbon dioxide in the atmosphere.

Precipitation, such as rain and snow, comes from water vapor in the air.

Carbon dioxide plays an important role in the lives of plants and algae. Carbon dioxide is removed from the air during **photosynthesis**. Plants and algae use light from the Sun, carbon dioxide, and water to produce sugar (food). During this process, they release oxygen to the atmosphere. When living organisms use the energy of food to stay alive, they remove oxygen from the air and return carbon dioxide to the air.

Plants make sugar out of sunlight, carbon dioxide, and water.

Here are other gases that you might have heard about. Methane is a variable gas that is increasing in concentration in the atmosphere. Scientists are trying to figure out why this is happening. They suspect several things. Cattle produce methane in their digestive processes. Methane also comes from coal mines, oil wells, and gas pipelines. It is a by-product of rice cultivation and melting permafrost in arctic regions. Methane **absorbs** energy and transfers heat to the atmosphere.

Ozone is a variable gas. It is a form of oxygen that forms a thin layer high in the atmosphere. The ozone layer protects life on Earth by absorbing dangerous ultraviolet (UV) light from the Sun. But ozone in high concentrations can cause lung damage. In the lower atmosphere, ozone is an air pollutant.

Some methane gas comes from the digestive processes of cows.

Thinking about Air

What is air?

Earth's Atmosphere

Earth's atmosphere is made up of a mixture of permanent and variable gases. These gases are all mixed together. Any sample of air is a mixture of all of them. The gases mix because the air particles are always moving near Earth's surface. Above about 90 kilometers (km), the gases mix less, and there are more light gases, such as hydrogen and helium.

There are extreme **temperatures** in the universe. The temperature can be as cold as –270 degrees Celsius (°C). Near hot stars, such as the Sun, it can be very hot, up to thousands of degrees. But there are a few places in the universe that have a temperature between those extremes of hot and cold. Earth is one of those places where the temperature is just right.

On a typical day, the temperature range on Earth is only about 100°C. It might be 45°C in the hottest place on Earth and –55°C at one of the poles. The measured extremes are 58°C in Al-Aziziyah, Libya, recorded on September 13, 1922, and –89°C in Vostok, Antarctica, on July 21, 1983. That's a range of temperature on Earth of 147°C.

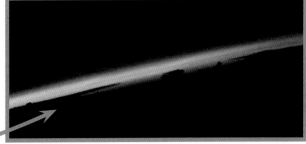

Space-shuttle astronauts took this photo while orbiting Earth. You can see a side view of Earth's atmosphere. The black bumps pushing into the troposphere are tall cumulus clouds.

The crew of Apollo 17 took this photo of Earth in December 1972, while on their way to the Moon. The small orange box at the top of Earth's image shows about how much area is in the atmosphere photo.

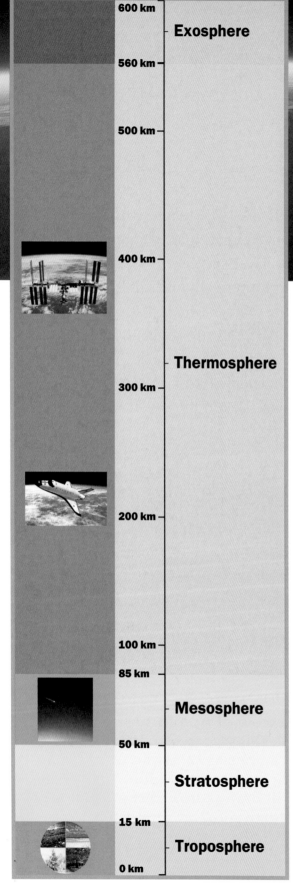

Earth's atmosphere looks like a thin, blue veil.

It's not only because we are at the right distance from the Sun that Earth's temperatures are moderate. Earth's atmosphere keeps the temperature within a narrow range so that it is just right for life on Earth.

From space, Earth's atmosphere looks like a thin, blue veil. Some people like to think of the atmosphere as an ocean of air covering Earth. The depth of this "ocean" is about 600 km. The atmosphere is most dense right at the bottom, where it rests on Earth's surface. It gets thinner and thinner (less dense) as you move away from Earth's surface. There is no real boundary between the atmosphere and space. The air just gets thinner and thinner until it disappears.

Imagine a column of air that starts on Earth's surface and extends up 600 km to the top of the atmosphere. Scientists have discovered several layers in this column of air. Each layer has a different temperature. Here's how it stacks up.

The four seasons occur in the troposphere.

The layer we live in is the **troposphere**. It starts at Earth's surface and extends upward for 9–20 km. Its thickness depends on the **season** and where you are on Earth. Over the warm equator, the troposphere is a little thicker than it is over the polar regions, where the air is colder. It also thickens during the summer and thins during the winter. The average thickness of the troposphere is 15 km.

This ground-floor layer has most of the organisms, dust, water vapor, and clouds found in the entire atmosphere. It has most of the air as well. And, most important, **weather** occurs in the troposphere. The troposphere is where differences in air temperature, **humidity**, **air pressure**, and **wind** occur.

These properties of temperature, humidity, air pressure, and wind are important **weather variables**. **Meteorologists** launch weather balloons twice each day to monitor weather variables. The balloons float up through the troposphere to about 18 km.

The troposphere is the thinnest layer. It has only about 2 percent of the depth of the atmosphere. It is the most dense layer, however, containing four-fifths (80 percent) of the total mass of the atmosphere.

Earth's surface (both land and water) absorbs heat from the Sun and warms the air above it. Because air in the troposphere is heated mostly by Earth's surface, the air is warmest close to the ground. The air temperature drops as you go higher. At its highest point, the temperature of the troposphere is about –60°C. The average temperature of the troposphere is about 25°C.

Mount Everest, located in Nepal and Tibet, is the highest landform on Earth, rising 8.848 km into the troposphere. The air temperature at the top of the mountain is far below freezing most of the time. There is also less air to breathe at the top of Mount Everest. Climbers usually bring oxygen to help them survive the thin air.

Mount Everest

Jets can fly in the region between the lower stratosphere and the upper troposphere.

The **stratosphere** is the layer above the troposphere. It is 15–50 km above Earth's surface and has almost no moisture or dust. It does, however, have a layer of ozone that absorbs ultraviolet (UV) radiation from the Sun. The temperature stays below freezing until you reach the top part of the stratosphere, where ozone absorbs energy and warms the air to about 0°C.

The jet stream, a fast-flowing stream of wind, travels generally west to east in the region between the lower stratosphere and the upper troposphere. Many military and commercial jets take advantage of the jet stream when flying from west to east. The jet stream winds move cold air over North America. This brings cold temperatures and storms.

The **mesosphere** is above the stratosphere, 50–85 km above Earth's surface. The temperature is colder than in the stratosphere. Its coldest temperature is around –90°C in the upper mesosphere. This is the layer where meteors burn up while entering Earth's atmosphere. We call these burning meteors shooting stars.

Beyond the mesosphere, 85–560 km (or higher) above Earth, is the **thermosphere**. The thermosphere is the most difficult layer of the atmosphere to measure. The air is extremely thin. The thermosphere is where the atmosphere is first heated by the Sun. A small amount of energy coming from the Sun can result in a large temperature change. When the Sun is very active with sunspots or flares, the temperature of the thermosphere can be 1,500°C or higher!

Temperature defines these four layers. The boundaries between the layers are not fixed lines and they can change with the seasons.

Meteors burn up in the mesosphere.

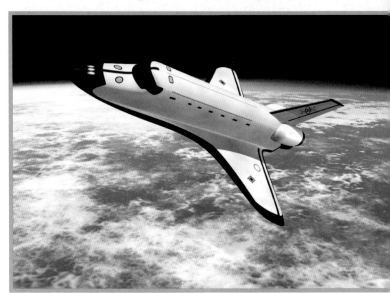

The space shuttle orbited Earth in the thermosphere.

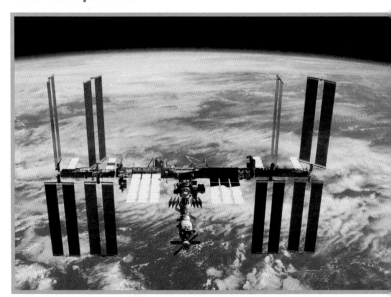

The International Space Station is in the thermosphere above Earth.

Earth's atmosphere fades into space.

Beyond the thermosphere, Earth's atmosphere fades into space. The **exosphere** is where gas particles escape into space. In this region, the temperature drops to the extreme –270°C of outer space.

That 600 km column of air pushes down on the surface of Earth with a lot of force. We call the force air pressure. We are not aware of it because we are adapted to live under all that pressure. But there is a mass of about 1 kilogram (kg) pushing down on every square centimeter (cm) of surface on Earth. Your head has a surface area of about 150 square cm. This means you have about 150 kg of air pushing down on your head. That's like having a kitchen stove or a motorcycle pushing down on your head all the time!

Here's another way to look at it. If all the air were replaced with solid gold, the entire planet would be covered by a layer of gold about half a meter deep. The mass of the entire atmosphere is about the same as half a meter of gold. But the atmosphere is much more valuable.

Thinking about the Atmosphere

1. How is Earth's atmosphere like the ocean? How is it unlike the ocean?
2. What is the average temperature of the troposphere? Why is that important?

Weather Instruments

A weather tower with a weather station on top

Meteorologists are scientists who study weather. Weather is the condition of the air in an area. The conditions can change, so they are called weather variables. The most important weather variables to meteorologists are temperature, air pressure, humidity, and wind. Meteorologists use weather instruments to measure each variable.

Temperature

Temperature is a measure of how hot the air is. Temperature is measured with a **thermometer**. There are many kinds of thermometers. The most common kind is a **liquid** thermometer. A liquid thermometer is a thin glass tube connected to a small bulb of liquid. As the liquid warms and cools, it expands and contracts. The height of the column of liquid in the tube changes in response to the temperature. By labeling the liquid tube to show temperatures, the meteorologist can read the temperature directly from the thermometer.

Metals also expand and contract in response to temperature change. Some thermometers use strips made of two different metals to detect temperature changes. These are called bimetallic thermometers. The two metals have different rates of expansion. One side of the strip expands more than the other as it heats up, and the strip bends. A pointer on the end of the bending strip points to the temperature.

A liquid thermometer

A bimetallic thermometer

cold   hot

14

Air Pressure

Air pressure is the force of air pushing on things around it. Air pressure changes with the density of the air. When air heats up, it becomes less dense; when it cools, it becomes more dense. The instrument that measures air pressure is called a **barometer**. Air pushes on a closed container, one side of which is attached to a dial in the barometer. The harder the air pushes, the higher the dial goes. The dial measures in units called millibars. Changes in air pressure mean that weather conditions will change. Falling air pressure means **rain** is coming. Rising air pressure means fair and dry weather is coming.

A barometer

Humidity

Water vapor is water (H_2O) in its gas state. As vapor, water can enter the air. The water vapor will eventually condense and form drops of water, which can fall as rain. Meteorologists measure humidity, the amount of water in the air, with instruments called **hygrometers**. Humidity is measured as a percentage.

A hygrometer

Wind Speed

Moving air is called wind. Meteorologists are interested in how fast the wind is moving. To measure wind speed, meteorologists use **anemometers** and **wind meters**. An anemometer uses a rotating shaft with wind-catching cups attached at the top. The harder the wind blows, the faster the shaft rotates, and the faster the cups move through the air. The moving cups measure the wind speed.

A wind meter is an instrument with a small ball in a tube. When wind blows across the top of the tube, the flow of air up the tube lifts the ball. The harder the wind blows, the higher the ball rises. Both instruments are adjusted to report wind in miles per hour (mph) or kilometers (km) per hour.

An anemometer

A wind meter

Wind Direction

Meteorologists are also interested in the direction the wind is blowing. To determine wind direction, meteorologists use a **wind vane**. A wind vane is a shaft with an arrow point on one end and a broad paddle shape at the other end. When wind hits the paddle, it rotates the shaft so that the arrow points into the wind. Using a compass, the meteorologist determines the direction the shaft is pointing. Wind direction is the direction from which the wind is blowing. It is reported in compass directions, such as north or south.

A wind vane

A compass

Modern Weather Instruments

Meteorologists now use a combination of traditional weather instruments and computer-based digital weather instruments. Meteorologists get information from advanced electronic instruments that are placed in good locations for monitoring weather. Those instruments use radio transmitters (like those in cell phones) to send information to weather centers where meteorologists work.

This weather device for home use has electronic instruments inside for detecting and reporting temperature and humidity. Some models measure air pressure and are connected to anemometers to measure wind speed.

Uneven Heating

Stars create a lot of energy. Energy radiates from them in all directions. Most of this energy streams out into space and never hits anything. A small amount, however, hits objects in the universe. When you look into the sky on a dark, clear night, you see thousands of stars. You see them because a tiny amount of energy from the stars goes into your eyes.

During the day, you are aware of the energy coming from a much closer star, the Sun. When sunlight comes to Earth, you can see the light and feel the heat when your skin absorbs the light. Heat and light from the Sun are called **solar energy**.

When light from the Sun hits matter, such as Earth's surface, two important things can happen. The light can be **reflected** or absorbed by the matter. If the light is reflected, it simply bounces off the matter and continues on its way in a new direction. But if the light is absorbed, the matter gains energy. Usually the gained energy is heat. When matter absorbs energy, its temperature goes up.

Heating It Up

The amount of solar energy coming from the Sun is about the same all the time. But the temperature of Earth's surface is not even. Some locations get warmer than other locations. Why is that?

There are several variables that affect how hot a material will get when solar energy shines on it. The table below lists several variables and how each affects the temperature change of a material.

Variable	Effect
Length of exposure	Longer exposure leads to higher temperature.
Intensity of solar energy	Greater intensity leads to higher temperature.
Angle of exposure	More direct angle leads to higher temperature.
Color of material	Darker color leads to higher temperature.
Properties of material	Water shows the least temperature change.

Length of exposure is how long the Sun shines on an object.

Intensity of solar energy is how bright and concentrated the energy is. For example, if the light travels through clouds, it will be less intense. Clouds reflect and absorb some of the energy before it gets to Earth's surface. The brighter the sunlight falling on an object, the warmer the object will get.

Angle of exposure changes throughout the day. Morning sunshine comes at a low angle and is less intense and weak. Midday sunshine radiates down from a high angle and is intense and strong.

Different colors absorb solar energy differently. Black absorbs the most solar energy. White absorbs the least solar energy.

The chemical properties of materials affect how hot they get when they absorb solar energy. Water heats up slowly and soil heats up rapidly when they absorb the same amount of energy. Water cools slowly and soil cools rapidly when they are moved to the shade.

Solar Energy in Action

Imagine a summer trip to the beach. On a cloudless day, the Sun shines down with equal intensity on the sandy beach and the ocean. It's a hot day.

When the car stops in the parking lot in the early afternoon, the pavement in the parking lot is hot. The black asphalt has absorbed a lot of solar energy, and its temperature is 50 degrees Celsius (°C). You walk across the parking lot (ouch, hot!) and onto the white sand. Whew! The white sand isn't as hot. It is a bearable 32°C. You keep moving toward the water. You finally get relief from the intense heat. The temperature of the ocean water is 22°C.

The asphalt, sand, and ocean water were all exposed to the same intensity of solar energy for the same length of time. But they are all different temperatures.

Black asphalt absorbs a lot of energy and gets very hot. White sand reflects a lot of solar energy. Light-colored sand doesn't get as hot as asphalt. Water absorbs a lot of energy, but it stays cool.

The temperature of Earth's surface is not the same everywhere. Land gets hotter than water in the sunlight. Land gets colder than water when the Sun goes down. Land heats up and cools off rapidly. Water heats up and cools off slowly. The result is uneven heating of Earth's surface.

Uneven Heating Worldwide

You can experience uneven heating of Earth's surface with your bare feet when walking on asphalt. The difference in temperature between the asphalt and water is obvious. On a larger scale, the whole planet is heated unevenly. The tropical areas (the tropics, near the equator) are warmer. The polar areas are cooler. That's because the angle of exposure of the solar energy is more direct in the tropics.

You can feel uneven heating of Earth when you are barefoot on asphalt.

The illustration shows how sunlight comes straight down on the tropics. But the sunlight comes at a sideways angle toward the poles. You can see how the same amount of light is spread over a much larger area in the north than in the tropics. This results in more heating in the tropics and less heating in the northern areas.

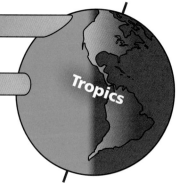

Two identical beams of sunlight. The upper beam spreads over a larger area toward the poles.

Thinking about Uneven Heating

1. What causes Earth's surface to heat up?
2. What are some of the variables that cause uneven heating of Earth's surface?
3. What happens to the temperature of equal volumes of soil and water when they are placed in the sunlight for 30 minutes?

Heating the Air: Radiation and Conduction

You might have had an experience like this one. The campfire has burned down to a bed of hot coals. Now it is time to toast some marshmallows. You put a marshmallow on a long stick and stand at a safe distance from the coals to toast your treat. You can feel the heat coming from the coals. After a minute, the marshmallow is brown and gooey. You pop it into your mouth. Yikes, that's hot! You didn't wait long enough for it to cool.

That story includes a couple of heat experiences. Have you ever stopped to think about what heat really is? What is the heat that you felt coming off the coals and the heat in the marshmallow that burned your tongue?

Heat = Movement

Objects in motion have energy. The faster they move, the more energy they have. Energy of motion is called **kinetic energy**.

Matter, like soccer balls, juice bottles, water, and air, is made of particles that are too small to see. The particles are in motion. They are in motion even in steel nails and glass bottles. In solids, the particles vibrate back and forth. In liquids and gases, the particles move all over the place. The faster the particles vibrate or move, the more energy they have.

Particle motion is kinetic energy, which can produce heat. The amount of kinetic energy in the particles of a material determines how much heat it produces. The particles in hot materials are moving fast. The particles in cold materials are moving more slowly.

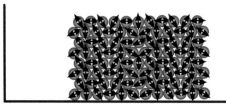

The particles in solids are held close to each other. They move by vibrating.

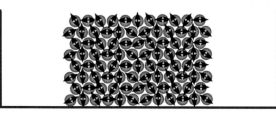

When solids get hot, the particles vibrate more. Hotter solids have more kinetic energy than colder solids.

The particles in liquids move by bumping and sliding around each other.

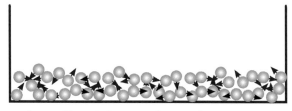

When liquids get hot, the particles bump and push each other more. Hotter liquids have more kinetic energy than colder liquids.

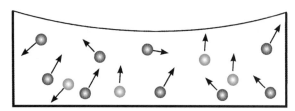

The particles in gases fly individually through the air in all directions.

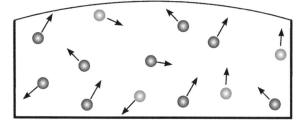

When gases get hot, the particles fly faster and are farther apart. Hotter gases have more kinetic energy than colder gases.

Energy Transfer

Heat can transfer (move) from one place to another. We can observe the **energy transfer** when energy is present as heat. Scientists sometimes describe heat transfer as heat flow, as though it were a liquid. Heat is not a liquid, but flow is a good way to imagine its movement.

Heat flows from a hotter location (high energy) to a cooler one (less energy). For example, if you add cold milk to your hot chocolate, heat flows from the hot chocolate to the cold milk. The hot chocolate cools because heat flows away. The cold milk warms because heat flows in. Soon the chocolate and the milk arrive at the same warm temperature, and you gulp them down.

Energy Transfer by Radiation

Burning gas, like the burner on a stove, can get very hot. When this happens, the burner is radiating heat and light. If you are close to a lightbulb, you can see light and feel heat, even though you are not touching the lightbulb. Energy that travels through air is **radiant energy**.

Radiant energy travels in rays. Heat and light rays radiate from sources like hot campfire coals, lightbulbs, and the Sun.

Radiation from the Sun passes through Earth's atmosphere. We call this solar energy. When solar energy hits a particle of matter, such as a gas particle in the air, a water particle in the ocean, or a particle of soil, the energy can be absorbed. Absorbed radiation increases the kinetic energy (movement) of the particles in the air, water, or soil. Increased kinetic energy produces a higher temperature, so the material gets hotter.

Radiation is one way energy moves from one place to another. Materials don't have to touch for energy to transfer from one place to the other.

Energy Transfer by Conduction

Imagine that hot toasted marshmallow or maybe a slice of pizza straight from the oven going into your mouth. This is another kind of energy transfer. When energy transfers from one place to another by contact, it is called **conduction**.

The fast-moving particles of the hot pizza bang into the slower particles of your mouth. The particles in your tongue gain kinetic energy. At the same time, particles of the hot pizza lose kinetic energy, so the pizza cools off. Some of the pizza's kinetic energy is conducted to heat receptors on your tongue, which sends a message to your brain that says, "Hot, hot, hot!"

When you heat water in a pot, the water gets hot because it touches the hot metal of the pot. Kinetic energy transfers from the hot metal particles to the cold water particles by contact, which is conduction.

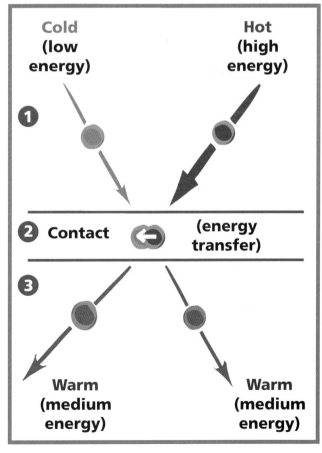

A "hot" particle with a lot of kinetic energy collides with a "cold" particle with little kinetic energy. Energy transfers at the point of contact. The cold particle gains energy, and the hot particle loses energy.

25

Energy Transfer to the Air

We have learned about energy transfer by radiation and conduction. Now let's explore energy from the Sun and what happens when it interacts with the air. Many kinds of rays radiate from the Sun. The most important rays are visible light and infrared light, which is invisible. So how do the Sun's rays heat up the air?

Air is 99 percent nitrogen and oxygen particles. Neither kind of particle absorbs visible light or infrared radiation. Only water vapor and carbon dioxide absorb significant amounts of radiant energy, and this is mostly infrared rays, not visible light.

If only a tiny part of the air absorbs the incoming radiant energy, how does the rest of the air get hot?

Earth's surface absorbs visible light. The land and water warm up. The air particles that touch the warm land and water particles gain energy by conduction. But there is more.

The warm land and water also radiate energy. This is a very important idea. Earth gives off infrared radiation that can be absorbed by water particles and carbon dioxide particles in the air. The energy transferred to the small number of water particles and carbon dioxide particles is transferred throughout the air by conduction. This happens when energized water particles and carbon dioxide particles bang into oxygen and nitrogen particles.

The air is not only heated from above. It is also heated from below.

Water particles warmed by radiation coming up from Earth

Air particles warmed by conduction

Earth warmed by solar energy radiates infrared radiation.

Wind and Convection

Wind lifts a kite high in the air.

Kite flying can be a lot of fun if the conditions are right. If the conditions are wrong, kite flying can be boring. What makes conditions right for kite flying? Wind.

Wind is air in motion. Air is matter. Air has mass and takes up space. When a mass of air is in motion, it can move things around. Wind can blow leaves down the street, lift your hat off your head, and carry a kite high in the air.

Sometimes air is still. Other times the wind is blowing. What causes the wind to blow? What puts the air into motion? The answer is energy. It takes energy to move air. The energy to create wind comes from the Sun.

Air is particles of nitrogen, oxygen, and a few other gases. The particles are flying around and banging into each other, the land, and the ocean. Let's imagine we are at the beach. It's early morning. The air over the land and the air over the ocean are both the same cool temperature.

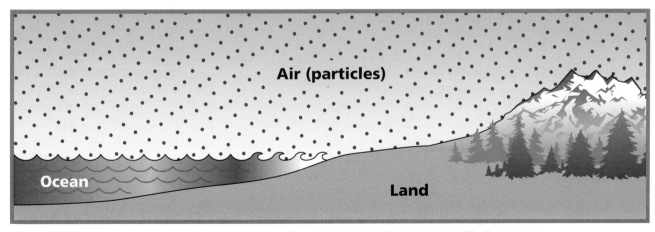

In the early morning, the land, ocean, and air are all the same cool temperature.

As the Sun shines down on the land and ocean, solar energy is absorbed. The land heats up quickly. The ocean heats up very slowly. By noon the land is hot, but the ocean is still cool. Earth's surface is heated unevenly. The afternoon wind starts. Here's why.

When air particles bang into the hot surface of the land, energy transfers to the air particles. Because of this energy transfer, the air particles fly around faster. The air gets hot. The hot-air particles bang into each other harder. That pushes the particles farther apart.

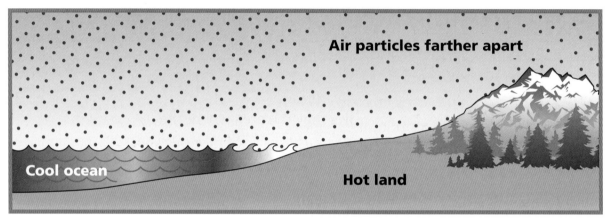

Energy transfers from the hot land to the air particles. The air particles move farther apart.

Over the ocean, air particles are banging into the cool surface of the water. The air stays cool. The air particles continue to move at a slower speed. The cool-air particles don't hit each other as hard, so they stay closer together.

A cubic meter of hot air has fewer particles than a cubic meter of cool air. The hot air is less dense than an equal volume of cool air.

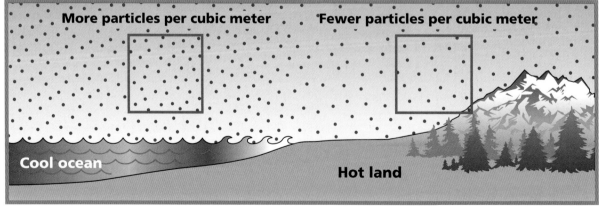

A cubic meter of hot air has fewer particles than a cubic meter of cold air. Hot air is less dense than cool air.

The Wind Starts

You know that cork floats on water. Cork floats on water because it is less dense than water. If you take a cork to the bottom of the ocean and let it go, it will float to the surface.

That's exactly what happens with warm and cold air. The warm air over the land floats upward because it is less dense than the cool air over the ocean. The more-dense, cool air flows into the area where the less-dense, warm air is and pushes it upward. The movement of more-dense air from the ocean to the warm land is wind. Wind is the movement of more-dense air to an area where the air is less dense.

Dense, cool air flows from the ocean to the land. Less-dense, warm air rises.

There is more to the story of wind. Two things happen at the same time to create wind. The warm air cools as it rises, becoming more dense than the surrounding air. At the same time, the dense air from the ocean warms up as it flows over the hot land.

As a result, air starts to move in a big circle. Air that is warmed by the hot land moves upward. The warm air cools as it moves up, gets more dense, and starts to fall. The rising and falling air sets up a big circular air current. The circular current is called a **convection current**.

As long as Earth's surface continues to be heated unevenly, the convection current will continue to flow. The part of the convection current that flows across Earth's surface is what we experience as wind. But what happens at night?

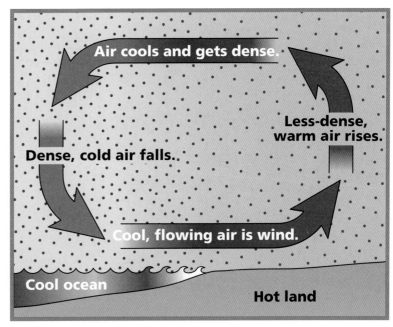

A convection current is the result of the uneven heating of Earth's surface.

The Wind Changes Direction

When the Sun goes down, solar energy no longer falls on the land and ocean. The land cools rapidly, but the ocean stays at about the same temperature. The air over the cool land is no longer heated. The density of the air over the land and ocean is the same. The convection current stops flowing. The wind stops blowing.

What will happen if the night is really cold? The land will get cold. The air over the land will get cold. The cold air will become more dense than the air over the ocean. The more-dense air will flow from the land to the ocean. The convection current will flow in the opposite direction, and the wind will blow from the land to the ocean.

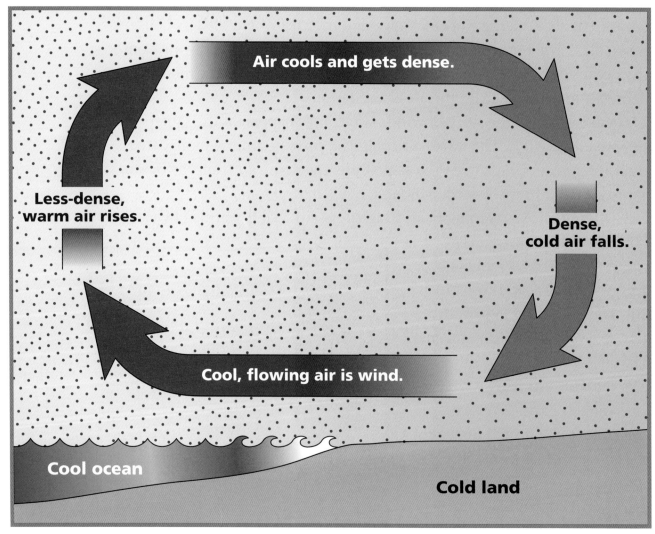

This convection current creates wind that blows from the land to the ocean.

Convection Summary

Uneven heating of Earth's surface by the Sun causes uneven heating of the air over Earth's surface. Uneven heating of air causes wind. Warm air is less dense than cold air. Cold, dense air flows to an area where the air is warmer and less dense. The less-dense air is pushed upward. As the warm air moves upward, it cools. Cool air is more dense, so it falls back to Earth. This circular pattern of air flow is a convection current. Convection currents are important ways that air masses move from place to place in the atmosphere. Convection currents transport energy from place to place.

Convection currents produce wind. The greater the difference in temperature between the warm and cold air masses, the harder and faster the wind will blow. Uneven heating of Earth's surface is the cause of many weather changes on Earth, including **hurricanes**, **tornadoes**, and **thunderstorms**.

Convection currents produce wind.

Uneven heating of Earth can cause hurricanes.

Thinking about Wind

1. How are convection currents produced in the air?
2. Explain what causes wind.
3. What happens to air particles when air is heated?
4. What is the source of energy that causes the wind to blow?

Wind Power

Wind energy has powered sailing ships for thousands of years.

Heat isn't the only energy resource that starts with the Sun. Wind, which is caused by the Sun's uneven heating of Earth, can also be used to generate power.

Wind is created when cool air rushes in to take the place of warmer, less-dense air. People have used wind power for thousands of years. In the past, sailing ships were the quickest and easiest way to travel. These ships had large sails to catch the wind and move them across the water.

One ancient use of wind power is windmills. Arabic people introduced windmills to Europe in about the 12th century. Ancient windmills were used to grind grain into flour. These windmills caught the wind in sails made of wood and cloth. The sails turned an axle, which transferred its motion to a turning pole. The pole turned the millstones that ground the grain. Windmills were also used to pump water.

Inventors and scientists have made many changes in windmills. Modern windmills are called wind turbines. A small motor starts the blades turning. Then they can turn on their own, even in wind speeds as low as 24 km per hour.

Windmills have provided power for pumping water and grinding grain in Europe since the 12th century.

Wind turbines convert wind energy into electricity.

Sometimes dozens or even hundreds of wind turbines are set up on a wind farm. One well-known wind farm is at Altamont Pass in California. As warm valley air rises in the central part of the state, cool air from the ocean flows over the coastal mountains. Narrow gaps between mountains, called passes, channel the air and increase its speed.

There are over 13,000 wind turbines in Altamont Pass and two other areas in California. In 1995, these three areas produced 30 percent of the world's wind-generated electricity. Roscoe Wind Farm in Texas is the largest land-based wind farm in the world. It has 627 wind turbines over 100,000 acres of west Texas. This wind farm can provide electricity to 230,000 homes. In 2010, wind power provided about 2 percent of the electricity generated in the United States.

Wind power has some disadvantages. The power produced depends on how strongly the wind is blowing. Therefore, wind farms can produce too much energy at some times and too little energy at others. Excess energy needs to be stored. It can be used to heat water or oil, or stored in batteries.

Some people don't like wind farms because they think the turbines are ugly. The turbines also create a whipping sound as they turn. The sound annoys some people and scares away animals. Birds can be killed by flying into spinning turbines.

Even though wind power has its problems, it is still an important source of energy. Each year, the United States uses the same amount of energy from wind turbines as 716,400 barrels of oil could provide. Using wind power instead of burning **fossil fuels** keeps about 2.5 million tons of carbon out of the air. Eventually fossil fuels will start to run out, and traditional methods of creating energy will become less desirable. Wind power, like solar power, is an energy source of the future.

Solar Technology

People use solar energy in a number of ways. People use it to heat water, to cook food, to warm and cool buildings, and to generate electricity.

Solar Water Heaters

Solar water heaters have been around for thousands of years. The ancient Romans created public baths supplied with water that flowed through heating channels. These channels were like canals, open to the Sun. The water channels were lined with black slate to absorb as much of the Sun's energy as possible. Centuries later, people painted metal tanks black and tilted them toward the Sun to warm the water inside.

In 1891, the first American commercial solar water heater, the Climax, was built. It was a black iron tank inside a wooden box. The box was lined with black felt and covered with glass. It sat on a roof exposed to the Sun.

The Climax worked fine on a sunny day. But what about cloudy days or nighttime? In 1909, the Day and Night water heating system was invented. This system included a solar heater and a separate insulated storage tank. After the Sun heated the water, it was piped into the storage tank where it remained warm.

Today, there are many types of solar water heaters, some of them shown on these pages.

A thermosyphon system solar water heater

A flat-plate collector solar water heater

One type of solar water heater available today uses a flat-plate solar collector in an insulated box. On top of the collector are small tubes filled with water to absorb the Sun's rays. The system is placed on the roof of a house. The Sun-heated water is piped down into an insulated storage tank in the house.

The many types of modern solar water heaters share some common features. They must be positioned to capture the Sun's energy so they are often on top of roofs. They are made of materials that absorb the heat energy and transfer the energy to water. The systems are insulated so the heat energy doesn't escape from the water. The water moves through pipes to a storage tank and is accessible by turning a faucet.

A glass tube collector solar water heater

Solar Cookers

Most of your meals are probably cooked in an oven or on a stove powered by electricity or natural gas. Electric microwave ovens are also popular for cooking food. But what about cooking where there is no access to gas or electricity? In developing countries, more than 2 billion people burn wood for cooking. But burning wood is bad for the environment and creates health risks for people. Large numbers of trees have to be cut down. Wood burning creates smoke that pollutes the air and causes breathing problems.

People can use solar cookers as a safe, inexpensive, pollution-free way to cook. A number of organizations have developed solar ovens in countries all over the world.

A solar cooker in the Zanskar Mountains of northern India

One of the most successful solar cooking projects is in Kenya. Professor Daniel Kammen, of the University of California, Berkeley, has introduced solar cookers to villages in Kenya. His simple solar ovens are boxes lined with reflective foil and covered with glass. First, he demonstrates how easy they are to use. Kammen and his coworkers mix up a stew or cake. They put it in the oven. Then Kammen explains how the oven works. He also explains how to build one. Finally he opens the oven and lets everyone eat some of the cooked food.

It takes about 3 hours to cook a pot of stew in a solar oven and about half an hour to boil a pot of water. Cooking food over a traditional fire takes less time.

A solar oven in Kenya

However, solar ovens eliminate the time-consuming chore of gathering firewood. Solar cooking also eliminates many respiratory illnesses, which are a common cause of death in Kenya.

If the community is interested in the solar oven, Kammen and his coworkers come back with plywood, foil, glass, and nails. They teach the villagers how to build their own ovens. So far, Kammen and his team have introduced hundreds of solar ovens in eastern Kenya.

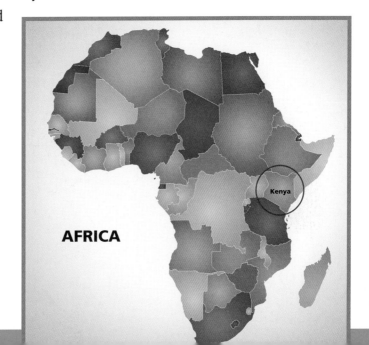

AFRICA

37

Solar Electricity

The Sun's energy can be converted into electricity. This process uses solar cells or steam-powered turbines. Solar cells are devices used to transform solar energy directly into electricity.

Solar cells are made of silicon, which is found in sand. The silicon is purified, crystallized, and cut into wafers or squares. Two wafers or squares are sandwiched together to form a solar cell. The cells are arranged on a flat panel made of glass, metal, or wood and sealed behind glass or plastic. One solar cell with a diameter of 10 centimeters (cm) can produce 1 watt of electric power. The electricity is then used immediately or stored in batteries.

Many solar cells are connected to make large panels called solar arrays. Solar arrays can generate enough electricity to store. The biggest challenge facing the use of solar cells is to bring down their cost. If this happens, many energy experts believe that solar cells could provide most of our energy needs by the end of this century.

Solar furnaces use huge mirrors to concentrate a lot of light in one place. Large solar furnaces can use this energy to produce steam, which can be used to turn large generators to produce large quantities of electricity. The solar furnace in Odeillo, France, has an 8-story-tall mirror and can produce temperatures up to 3,000 degrees Celsius (°C). That's hot enough to melt a steel plate in just 3.5 seconds!

The solar furnace in Odeillo, France

Infinium, after winning the American Solar Challenge in 2010

Solar energy is being studied as a source of energy for transportation. College engineering students from all over the country take part in a solar car race called the American Solar Challenge. The object of this event is to design, build, test, and race a solar-powered car in a cross-country race. The teams travel 1,760 kilometers (km) from Broken Arrow, Oklahoma, to Naperville, Illinois.

The cars taking part in the 2010 American Solar Challenge didn't look like anything parked in your driveway at home. The winner that year was University of Michigan's entry, called Infinium. Over the course of six days, Infinium traveled at an average speed of 64 km per hour. Infinium crossed the finish line with a time of 28 hours, 14 minutes, and 44 seconds.

Maria Telkes

Dr. Maria Telkes (1900–1995) was the world's most famous female inventor in the field of solar energy. Some people even called her the Sun Queen!

Maria Telkes was born in Hungary. She lived most of her life in the United States. She first became interested in solar energy when she was with the Massachusetts Institute of Technology (MIT) in the 1940s. During her 50-year career, she worked for many top universities.

Dr. Maria Telkes

Solar cooking was of special interest to Telkes. She realized that we need to **conserve** fuel. The heat of the Sun provided a clean source of energy. During the 1950s, she invented a model of a solar oven that is still used today. Later the Ford Foundation gave Telkes a $45,000 grant to work on her solar oven. This allowed Telkes to improve her design.

Telkes also helped to create a solar house. In 1981, she worked to design and build the Carlisle House in Carlisle, Massachusetts. MIT and the US Department of Energy also worked on the project. This house features a solar heating and cooling system. It uses no fossil fuels, such as natural gas, oil, and coal. The house generates so much power that it is able to share its extra energy with the local utility company.

The solar-powered Carlisle House was built by Telkes in 1981.

Solar Buildings

Imagine a house or office building where the electric, heating, and cooling systems don't rely on oil, natural gas, or coal. Instead, all these systems use energy from the Sun for power. Homes like this do exist. They are called solar houses.

Solar houses have many different systems that use energy from the Sun. Some of these systems are passive. They don't use any mechanical devices such as pumps or generators. An example of passive space heating would use large windows that face the Sun. The windows let sunlight shine in to warm the house. When the weather is warm, window coverings come down over the windows and keep the house cool. Solar houses are designed to work with the Sun to maintain a comfortable temperature at all times of the year.

Other solar systems are active. They use mechanical devices, such as fans and pumps, to move captured solar energy throughout the house. For example, many solar houses use flat-plate collectors to heat water and air. A flat-plate collector sits in a box that is insulated on the bottom and sides. The top is covered by one or more layers of clear glass or plastic. The system is placed on the roof of a house. The visible light goes through the glass, where it is absorbed by the plate and converted into heat. The glass traps the heat inside the box. The hot air is pumped to rooms in the house where it heats the space.

Solar energy is good for the environment! That's why some houses and apartment buildings use solar panels.

Solar energy can also be used to cool buildings during the summer. Most solar air-conditioning systems use solar collectors and materials called desiccants. Desiccants can absorb large amounts of water. Fans force air from outdoors through the desiccants, which remove moisture from the air. Next the dry air flows through a heat exchanger that removes some of the heat. Then the air passes over a surface soaked with water. As the water comes in contact with the dry air, it evaporates and removes more heat from the air. Finally the cooled air is pumped throughout the building.

Other types of active systems involve solar cells. Solar arrays (also called solar panels) can generate electricity for all kinds of household uses. Solar panels can be placed on the roof of a house or on the sides of apartment buildings. It is important that the solar cells capture as much sunlight during the day as possible.

Solar houses have many advantages. They are environmentally friendly. Solar energy is a **renewable resource**. It does not use up Earth's resources the way burning fossil fuels or wood does. In addition, solar collectors are pollution-free. They don't create fumes or other dangerous chemicals that can poison the environment or make people sick.

Thinking about Solar Technology

1. What are some features of solar water heaters used today?
2. What are the advantages of solar cookers?
3. What did Maria Telkes contribute to solar technology, and when did she do this?

Condensation

When water evaporates, where does it go? It goes into the air. Water is always evaporating. Clothes are drying on clotheslines. Wet streets are drying after a rain. Water is evaporating from lakes and the ocean all the time. Every day more than 1,000 cubic kilometers (km) of water evaporates worldwide. And all that water vapor goes into the air! That amount of water would cover the entire state of California 3 meters (m) deep.

What happens to all that water in the air? As long as the air stays warm, the water stays in the air as water vapor. Warmth (heat) indicates the presence of energy. As long as the particles of water vapor have a lot of kinetic energy, they continue to exist as gas.

But if the air cools, things change. As the air cools, the particles lose kinetic energy and slow down. When this happens, particles of water vapor start to come together. Slowing down and coming together is called condensation. Condensation is the change from gas to liquid.

Particles of condensed water vapor form tiny masses (droplets) of liquid water. When invisible water vapor in the atmosphere condenses, the water becomes visible again. Clouds and **fog** are made of these tiny droplets of liquid water.

The clouds in the sky are made of tiny droplets of water.

Condensation usually happens on a cold surface. In class, you observed condensation on the cold surface of a plastic cup filled with ice water. But there are no cups of ice water in the sky. What kind of surface does water vapor condense on?

Most condensation in the air starts with dust particles. Water particles attach to a dust particle. When a tiny mass of water has attached to a dust particle, other water particles will join the liquid mass.

When you look up in the sky and see clouds, you are seeing trillions of droplets of liquid water. Each droplet is made up of billions of water particles, but a single droplet is still too small to see. You can see them when trillions and trillions of them are close together in clouds.

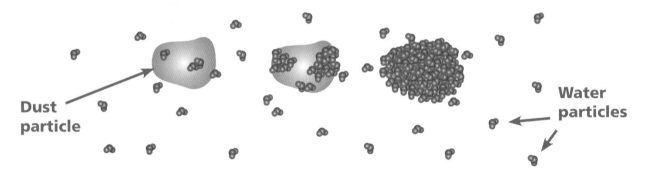

The mass of water grows and grows until it forms a tiny droplet of water.

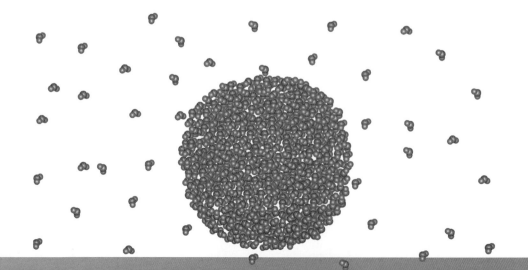

Fog close to the ground

Dew on a spider's web

Where else have you seen condensation other than up in the clouds? Sometimes water vapor condenses close to the ground. This is called fog. Being in fog is really being in a cloud that is at ground level.

Water vapor doesn't always condense in air. If you go out early in the morning after a warm day, you might see condensation called **dew**. In these pictures, dew formed on a spider's web and on a flower.

Water vapor condenses indoors, too. On a cold morning you might see condensation on your kitchen window. Or if you go outside into the cold wearing your glasses, they could get fogged with condensation when you go back inside.

Dew on a flower

Condensation on a window

What happens to the bathroom mirror after you take a shower? The air in the bathroom is warm and full of water vapor. When the air makes contact with the cool mirror, the water vapor condenses on the smooth surface. That's why the mirror is foggy and wet.

When the temperature drops below the freezing point of water (0°C), water vapor will condense and freeze. Frozen condensation is called **frost**. Frost is tiny crystals of ice. Frost might form on a car window on a cold night. You can also see frost on plants early on a winter morning. But you have to get up before the Sun if you want to see the beautiful frost patterns.

Condensation on a mirror

Frost on a window

Frost on grass and a leaf

Thinking about Condensation

1. What is condensation?
2. What role does temperature play in condensation?
3. What is frost?
4. Why does condensation form on a glass of iced tea?

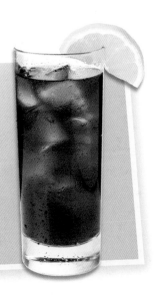

Where Is Earth's Water?

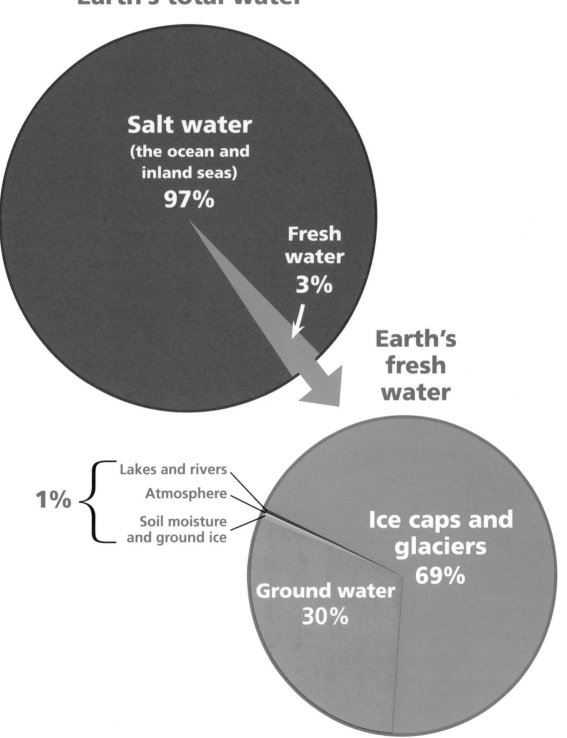

47

The Mississippi River

The Water Cycle

Water particles in the water you drink today might have once flowed down the Mississippi River. Those same particles might have washed one of George Washington's shirts. They might even have been in a puddle lapped up by a thirsty dinosaur a million years ago!

Water is in constant motion on Earth. You can see water in motion in rushing streams, falling raindrops, and blowing snowflakes. But water is in motion in other places, too. Water is flowing slowly through the soil. Water is drifting across the sky in clouds. Water is rising through the roots and stems of plants. Water is in motion all over the world.

Think about the Mississippi River for a moment. It flows all year long, year after year. Where does the water come from to keep the river flowing?

The water flowing in the river is renewed all the time. Rain and **snow** fall in and near the Mississippi River. Rain falling nearby soaks into the soil and runs into the river. The snow melts in the spring and supplies water for the river during the summer. Rain and snow keep the Mississippi River flowing.

The rain and snow in the Mississippi River are just a tiny part of a global system of water recycling. The system is called the **water cycle**.

The big idea of the water cycle is this. Water evaporates from Earth's surface and goes into the atmosphere. Water in the atmosphere moves to a new location. The water then returns to Earth's surface in the new location. The new location gets a new supply of water.

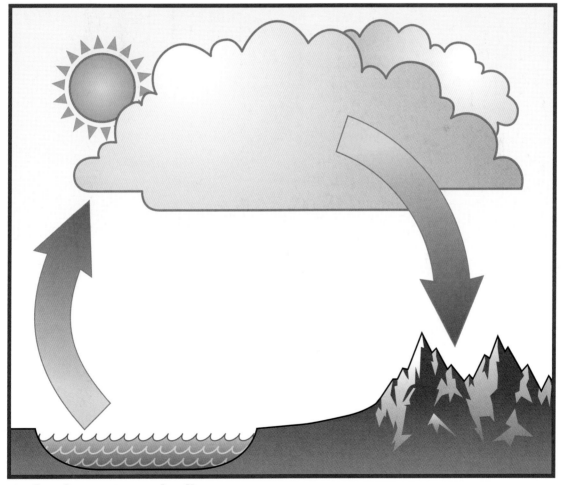

A simple water-cycle diagram

Water Evaporates from Earth's Surface

The Sun drives the water cycle. Energy from the Sun falls on Earth's surface and changes liquid water into water vapor. The ocean is where most of the evaporation takes place. But water evaporates from lakes, rivers, soil, wet city streets, plants, animals, and wherever there is water. Water evaporates from all parts of Earth's surface, both water and land.

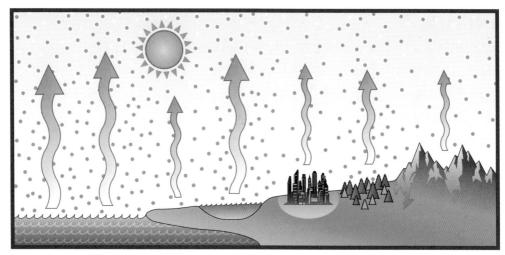

Water evaporates from all of Earth's surfaces.

Water vapor is made of individual water particles. Water vapor enters the air and makes it moist. The moist air moves up in the atmosphere. As moist air rises, it cools. When water vapor cools, it condenses. Water in the atmosphere changes from gas to liquid. Tiny droplets of liquid water form. The condensed water is visible as clouds, fog, and dew.

Water vapor condenses in the atmosphere to form clouds.

Water Falls Back to Earth's Surface

Wind blows clouds around. Clouds end up over mountains, forests, cities, deserts, and the ocean. When clouds are full of condensed water, the water falls back to Earth's surface as rain. If the temperature is really cold, the water will freeze and fall to Earth's surface as snow, **sleet**, or **hail**.

Water falls back to Earth's surface as rain, snow, sleet, or hail.

Water particles move through the water cycle at different speeds. They also follow different paths. For example, rain might soak into the soil. A particle might be taken in by plant roots. It might soon escape into the air through holes in plant leaves during a process called **transpiration**. If the air is cool, water might condense immediately as dew and fall back onto the soil. This is a very small water cycle that **recycles** water back to its starting place quickly.

Rain that lands on the roof of your school might flow to the ground. From there it could enter a stream. After a long journey, it could find its way to the ocean. There the rainwater could reenter the atmosphere as water vapor. By the time it condenses with millions of other particles to form a drop, the rainwater could be hundreds of kilometers (km) away from where it started. When the particle returns to Earth's surface, it could fall on the roof of a school in another state. This is an example of a large water cycle that moves water to a new location.

Rain can sink into the ground or freeze in a glacier. A particle far underground or deep in a mass of ice can take a long time to reenter the water cycle. It might take 100 years for a particle of ground water to come to the surface in a spring, and even longer for a particle to break free from a glacier.

The Sun provides the energy to change liquid water into vapor. Water vapor enters the air, where it is carried around the world. When water condenses, gravity pulls it back to Earth's surface. That's the water cycle, and it goes on endlessly.

Thinking about the Water Cycle

1. What is the water cycle?
2. When water falls from clouds, what forms can it take?
3. Describe a large water cycle that takes a long time to complete.
4. Describe a small water cycle that takes a short time to complete.

Severe Weather

Hurricane Katrina making landfall on the Gulf Coast

On August 29, 2005, Hurricane Katrina roared across the Gulf of Mexico and onto land. Throughout the country, people watched TV and listened to the radio as Katrina plowed into the states of Louisiana, Mississippi, and Alabama. The wind speed was 250 kilometers (km) per hour. The rain poured down. When the storm had passed, hundreds of people were dead, hundreds of thousands were homeless, and the city of New Orleans was flooded. The cost of the damage was in the billions of dollars.

Weather is fairly predictable most of the time. During the summer months in San Francisco, California, mornings and afternoons are often foggy. There might be sunshine in the middle of the day. In the winter months, rain is common. In Los Angeles, California, hot, dry weather is typical in the summer. In Gulf states, summer days are often hot and humid. In the Midwest and East, winters are usually cold, cloudy, and snowy. These are the normal weather conditions that people come to expect where they live.

It's the change from normal to the extreme that catches people's attention. Tornadoes, thunderstorms, windstorms, hurricanes, **droughts**, and floods are examples of **severe weather**. Severe weather brings out-of-the-ordinary conditions. It may cause dangerous situations that can damage property and threaten lives.

Rain is a common type of precipitation.

What Is Weather?

We are surrounded by air. It's a little bit like living on the bottom of an ocean of air. Things are always going on in the air surrounding us. The condition of the air around us is what we call weather.

A sunny day in Chicago, Illinois

Weather can be described in terms of four important variables. They are temperature, humidity, air pressure, and wind. They are called variables because they change. A day with nice weather might be warm, but not too hot. The sky is clear with just a little bit of moisture in the air. The air is still or moving with a light breeze. That's a perfect day for most people. But not too many days are perfect. Usually it's too hot, too humid, too windy, or too something. But don't worry. Weather always changes.

What Causes Weather to Change?

Energy makes weather happen. Energy makes weather change. The source of energy to create and change weather is the Sun.

When sunlight is intense, the air gets hot. When sunlight is blocked by clouds, or when the Sun goes down, the air cools off.

Moisture in the air takes the form of humidity, clouds, and precipitation. Intense sunlight evaporates more water from the land and ocean of Earth's surface. The result is more humidity, more cloud formation, and more rain. When sunlight is less intense, evaporation slows down.

Movement of air is wind. Uneven heating of Earth's surface results in uneven heating of the air touching Earth's surface. Warm air expands and gets less dense. More-dense, cool air flows under the warm air. This starts a convection current. The air flowing from the cool surface to the warm surface is wind.

When air pressure falls, rain is likely. A storm is possible.

Stormy weather approaching

Hurricane Earl near the Caribbean Islands in 2010

Hurricanes and Tropical Storms

Hurricanes are wind systems that rotate around an eye, or center of low air pressure. Hurricanes form over warm tropical seas. They are classified on a scale from 1 to 5, with 5 being the most powerful storm. Katrina was category 4 as it approached the Gulf Coast of the United States.

Most hurricanes that hit the United States start as tropical storms in the Atlantic Ocean. They form during late summer and early fall when the ocean is warmest. As a tropical storm moves west, it draws energy from the warm ocean water. The storm gets larger and stronger, and the wind spins faster and faster.

The spinning wind draws a lot of warm water vapor high in the storm system. When the vapor cools, it condenses. Condensation releases even more energy, which makes the system spin even faster. When the hurricane reaches land, the winds are blowing at deadly speeds, up to 250 km per hour. The rain is very heavy. The wind and rain can cause a lot of destruction.

As soon as a hurricane moves over land, it begins to lose strength. It no longer has warm water to give it energy and water vapor. Within hours, the wind and rain drop to safe levels.

Thunderstorms

Thunderstorms form when an air mass at the ground is much warmer and more humid than the air above. Rapid convection begins. As the warm, humid air rises, the water vapor in it condenses. The condensing water vapor transfers energy to the surrounding air, causing the air to rise even higher. The rapid movement of air also creates a static electric charge in the clouds. When the static electricity discharges, lightning travels from the clouds to the ground, and you hear the sound of thunder. Thunderstorms can cause death, start fires, and destroy communications systems. The powerful winds and heavy rain can cause property damage.

Thunderstorms are most common over land during the afternoon. The Sun heats Earth's surface, and heat transfers to the air. When cold air flows under the warm, moist air, thunderstorms are possible.

Lightning travels from the clouds to the ground.

A tornado spinning through a city

Tornadoes

Tornadoes are powerful forms of wind. They most often happen in late afternoons in spring or summer. When cold air over the land runs into a mass of warm air, the warm air is forced upward violently. At the same time, cooler, more-dense air flows in from the sides and twists the rising warm air. A spinning funnel forms. It "sucks up" everything in its path like a giant vacuum cleaner. The air pressure inside the funnel is very low. The air pressure outside the funnel is much higher. The extreme difference in air pressure can create wind speeds of 400 km per hour or more. Tornadoes can seriously damage everything in their path.

Tornadoes are most common in the south central part of the United States, from Texas to Nebraska. Hundreds of tornadoes occur in this region each year. Warm, moist air from the Gulf of Mexico moves northward. It runs into cooler, drier air flowing down from Canada. This creates perfect conditions for tornadoes. That's why this part of the United States is called Tornado Alley.

A tornado over water is called a waterspout.

Hot and Cold

Hot and cold weather are the direct result of solar energy. It gets hot when energy from the Sun is intense. It gets cold when solar energy is low. The ocean also affects temperature. The highest and lowest temperatures are never close to the ocean. Water has the ability to absorb and release large amounts of energy without changing temperature much. This keeps places close to the ocean from getting really hot or really cold.

Death Valley is one of the hottest places on Earth.

Here is a table of temperature extremes for the United States and the world. These temperatures are deadly for most organisms. Only a few tough organisms are able to survive such temperatures.

Area	Location	High Temperature	Low Temperature
United States	Death Valley, California	57°C	
	Prospect Creek, Alaska		−62°C
World	Al-Aziziyah, Libya	58°C	
	Vostok, Antarctica		−89°C

Weather Extremes

The West Coast and Northeast region of North America do not have many hurricanes or tornadoes. But they do have weather extremes. Most of them involve the ocean.

During the winter, it often rains and snows along the East and West Coasts and in the western mountains. When large storms come in from the Atlantic or Pacific Ocean, wind and rain can cause property damage and flooding. In the mountains, the precipitation comes down as snow. Intense snowstorms are called **blizzards**. A single blizzard can drop 4 meters (m) or more of snow. The snow for a whole winter might exceed 10 m.

A blizzard can drop more than 4 meters (m) of snow.

An ice storm can cause a lot of damage.

The Pineapple Express is a band of warm, moist air that flows to the West Coast from the warm ocean around the Hawaiian Islands. When the warm, humid Pineapple Express meets cold air flowing down from Alaska, a violent winter storm can develop. High winds and heavy rain can uproot trees, destroy homes, and flood large areas of lowlands.

When seasonal rain and snow fail to develop, droughts can occur. A drought is less-than-normal precipitation. In the Southwest, this means less rain in the deserts and hills, and less snow in the mountains. Less snow means less spring runoff. Less runoff means less flow in rivers and streams. Lakes and ponds shrink and in some cases dry up completely. Soil moisture dries up, and ground water decreases. Reservoirs that people use to store water shrink.

The Pineapple Express carries large amounts of moisture to California.

Water from lakes and rivers dries up during a drought.

Droughts put stress on natural and human communities. Fish and other aquatic organisms might die. Plants that are not adapted for dry environments might die. Reduced water for crops means less food production. People have to conserve water by using less and recycling water when possible.

Serious droughts are not uncommon. During the early 1930s, parts of Colorado, Kansas, New Mexico, Oklahoma, and Texas received little rain. Crops failed. Then came the strong winds. The farms in the area were stripped of their rich topsoil. The farmers had to leave the area because their fields were destroyed. Thousands of families had to leave the area known as the Dust Bowl.

Could it happen again? Many climate scientists think it is happening again now. The precipitation in the Southwest has been declining since the early years of this century. Stream flow and ground water are reduced. Reservoirs are low. The drought that has settled over the Southwest could be part of the overall change in the worldwide **climate**. People in the Southwest should be prepared to use less water. And they should be aware that a general drying of the land could result in more and hotter wildfires.

The Role of the Ocean in Weather

The ocean affects weather in the United States in several ways. The ocean is the source of most of the precipitation that falls on the West Coast states. Water evaporates from the ocean, particularly where the Sun has warmed the ocean's surface. Wind carries the water vapor and clouds over the land. As the moist air rises and cools over the coastal mountains, the Sierra Nevada, and the Cascade Range, the water vapor condenses and falls back to Earth's surface. During the spring and summer, the water flows back to the ocean, to complete the water cycle.

The ocean affects the weather.

The ocean creates mild temperatures all year along the West Coast. It rarely gets too hot or too cold. The temperature of the ocean doesn't change quickly. So the ocean keeps the air temperature near the coast even all year.

The ocean creates breezes near the coast. Because water heats up and cools down slowly, there is often a difference in the temperature of the land and the ocean. Uneven heating starts a convection current, which results in wind. The Sun and the ocean are responsible for ocean breezes.

Thinking about Severe Weather

1. What causes hurricanes?
2. What causes tornadoes?
3. How does the water cycle affect weather along the West Coast?
4. How does the ocean influence the weather along the West Coast?

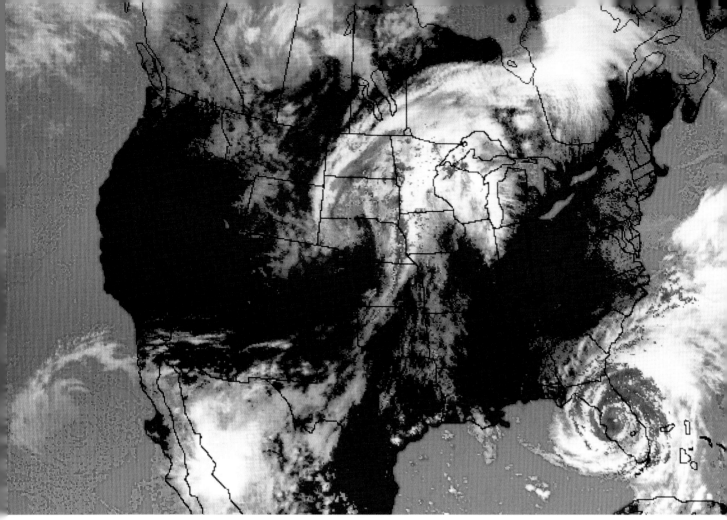

A weather satellite image enhanced with color

Weather Maps

Meteorologists collect information about the condition of the atmosphere. They measure air temperature, humidity, and air pressure. They measure wind speed and precipitation. They keep track of the movements of masses of warm air and cold air.

But meteorologists want to know more than what the weather is today. They want to **predict** what the weather will be like tomorrow and the day after that. Predicting weather is called **forecasting**.

Meteorologists collect information from surface measurements, atmospheric measurements, and satellite images. They analyze information from all three sources to make a forecast.

Surface Measurements

Weather data are collected every hour at more than 300 stations across the United States. At these locations, meteorologists measure several weather variables, including temperature, wind speed and direction, air pressure, cloud cover, and precipitation. These data are fed into weather-service computers. The computers generate surface-weather maps.

The surface-weather map has a code at each measuring station. The code is a combination of numbers and symbols. Information about all the weather variables can be read for each station.

Here's how some of the information is coded.

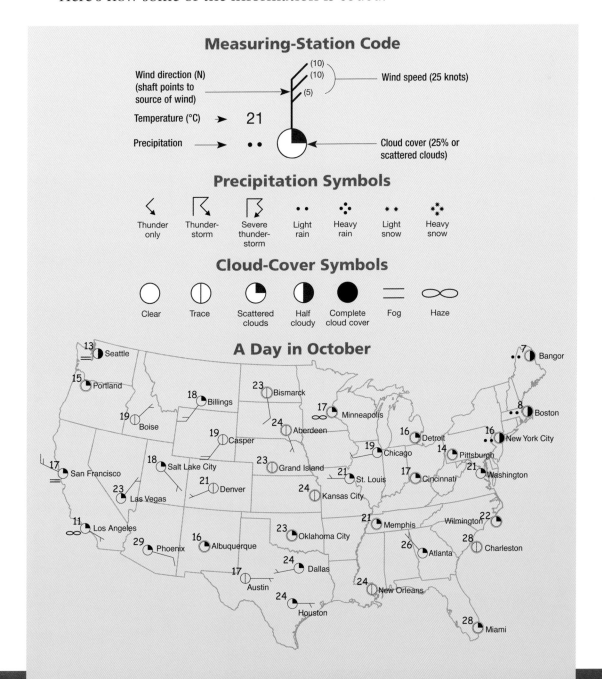

Atmospheric Measurements

Weather balloons carry instruments into the upper atmosphere to collect data twice each day. The balloons are released at exactly the same time all over the world. There are 93 release stations in the United States. In Washington state, the balloons go up at 4:00 a.m. and 4:00 p.m. In Pennsylvania, the balloons go up at 7:00 a.m. and 7:00 p.m.

The balloons carry **radiosondes**, instruments that measure air temperature, air pressure, and humidity. The radiosonde sends information to the station until the balloon pops. Meteorologists also track the balloon's path to figure out wind speed and wind direction.

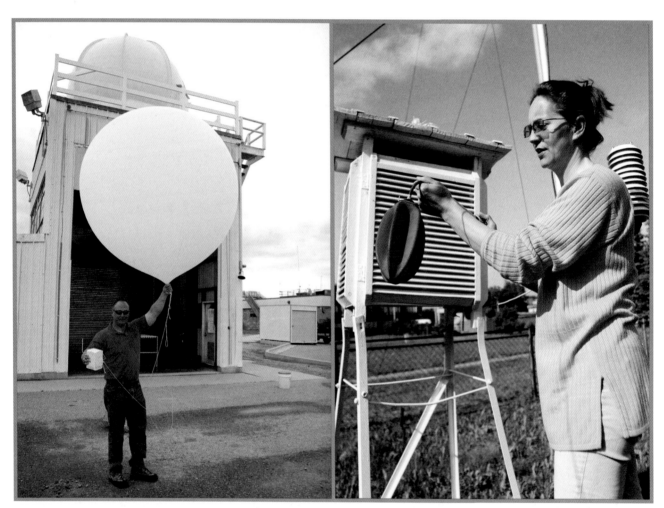

Meteorologists use weather balloons to collect weather data twice each day.

Satellite Images

Earth is surrounded by satellites stationed about 35,000 kilometers (km) above Earth's surface. The satellites are "geostationary." That means they orbit the rotating Earth at a speed that keeps them directly above the same point on Earth's surface at all times. The satellite appears to be stationary, or not moving, with respect to a fixed point on Earth. These satellites "watch" the clouds and water vapor move over Earth's surface. They read the temperature of Earth's surface. They identify the storm centers. All this information is sent back to Earth. Meteorologists use powerful computers to change the signals from the satellites into images of Earth's weather.

Satellites monitor Earth's weather.

Illustration of GOES weather satellite: Geostationary Operational Environment Satellite

Making Weather Maps

Meteorologists bring together the information from surface measurements, atmospheric measurements, and satellite images. Then they make **weather maps**. A weather map is a way to show weather data as a picture. A basic weather map is a picture of high pressure and low pressure, temperatures, and places where masses of warm air and cold air meet, called **fronts**. With this information on a map, a meteorologist can make a good weather forecast.

Reading Weather Maps

The Sun heats Earth's surface more near the equator than at the poles. Huge masses of air over the Caribbean Sea become warm. At the same time, huge masses of air in Alaska and northern Canada become cold. The masses of cold air move south, and the masses of warm air move north. When they meet, the area of contact is called a front. Weather changes happen at fronts.

A **cold front** happens when a cold air mass collides with a mass of warm, moist air. When this happens, the cold air pushes under the warm air and pushes it into the upper atmosphere very rapidly. The warm air cools, water condenses, and a thunderstorm occurs, producing a downpour. If the temperature difference between the air masses is large, a tornado might develop.

A cold air mass plows under a warm air mass, pushing it high in the atmosphere. Heavy rain and lightning occur for a short time.

A **warm front** happens when a warm air mass overtakes a cold air mass. The warm air slides over the top of the cold air in a long, slanting wedge. The warm air rises and cools slowly, and water vapor condenses into liquid over a long time. Warm fronts produce light rain for a long time.

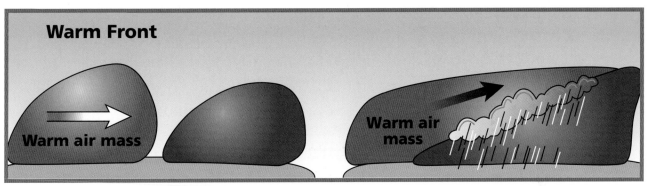

A warm air mass slides over a cold air mass. The warm air cools and produces light rain for a long time.

On a weather map, a line with points shows where a cold front is. The points show which direction the cold front is moving. A line with round bumps is used to show where a warm front is. The side of the line with the bumps is the direction the warm front is moving. When the map is prepared in color, cold fronts are blue and warm fronts are red.

Sometimes a warm front and a cold front come together and stop moving. This is called a **stationary front**. It is shown by a line with points on one side and bumps on the other. The weather under a stationary front is similar to the weather produced by a cold front.

67

On a weather map, high-pressure areas are shown with a large letter H. Low-pressure areas are shown with a large letter L.

The weather around a high-pressure center is usually cool and dry. That's because high pressure is associated with more-dense air. More-dense air tends to be cool and dry.

Low-pressure areas are usually warmer and moist. That's because low pressure is associated with less-dense air.

When a low-pressure area is near a high-pressure area, air will move from the high-pressure area to the low-pressure area. The movement of air is wind. The weather around a low-pressure area is windy and possibly rainy. As the warm air rises, cools, and condenses into clouds, it could start to rain.

Look at the three weather maps. Notice the large cold front going from Texas to New York. The upper part of the front moves across several states in the East. The southern end of the front, however, is stationary.

In Map 1, a cold front meets a warm front in Canada. Warm, moist air rises and condenses. The forecast is for rain.

In Map 2, a low-pressure area developed over North Dakota and South Dakota Monday afternoon. Air from the high-pressure area over the Rocky Mountains in Colorado might flow across Wyoming and Nebraska to the low-pressure area. The forecast is for wind.

Look at Map 3. What is the weather in California? There is a low-pressure area off the coast of Southern California. Cool, moist air from the ocean is flowing toward the low-pressure area. If the moist air warms and rises as it approaches the warm, low-pressure area, it could cool and condense. The forecast is for clouds and possible showers.

Look in the lower right-hand corner of each map. There is a red symbol for a tropical storm or hurricane. This is a hurricane. It is traveling past the southern tip of Florida. The forecast is for extreme wind and rain.

H	high-pressure area
L	low-pressure area
🌀	tropical storm or hurricane

Map 1:
Monday morning at 8:00 a.m.

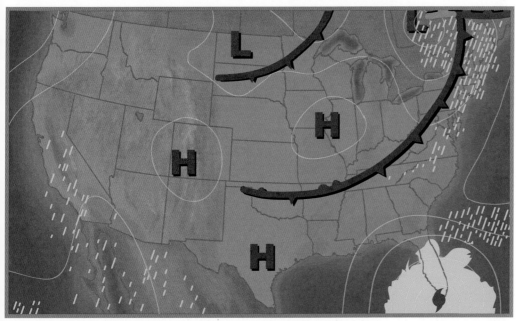

Map 2:
Monday afternoon at 2:00 p.m.

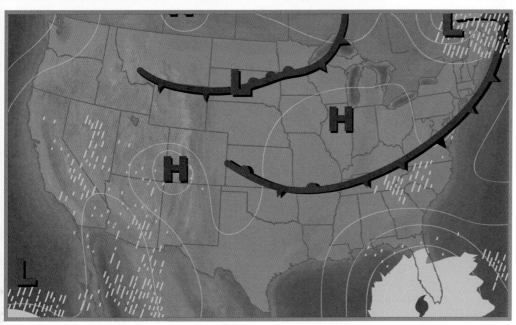

Map 3:
Monday evening at 8:00 p.m.

69

At National Weather Service forecast offices all around the United States, meteorologists use their skill and experience to produce weather forecasts. They consider all the weather variables such as temperature, air pressure, humidity, and wind. And they use their knowledge of general weather patterns.

Meteorologists know that winds in the upper atmosphere blow from west to east over most of the United States. So they know that most big weather systems also move from west to east.

They also know that air flows from high-pressure areas to low-pressure areas. This creates wind. So they look for high-pressure and low-pressure areas on the weather map to figure out which way and how hard the wind will blow.

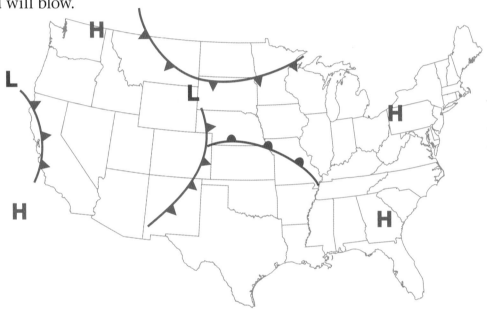

Thinking about Weather Maps

1. What weather variables do meteorologists measure when they are preparing to make a weather forecast?

2. Describe three kinds of fronts and the weather they produce.

3. What causes wind?

4. Look at the sample weather map above. Where do you think it is raining? Where will it be raining tomorrow?

5. Look at the high-pressure and low-pressure centers on the map above. Where do you think the wind is blowing? What direction?

6. Use the map above to predict where it is likely to be cold and dry.

Earth's Climates

What's the weather like today? What was it like last year on this same date? Probably just about the same. We can guess what the weather will be like tomorrow and next year at this time because weather tends to follow predictable patterns over long periods of time. The big patterns of weather define a region's climate. Climate describes the average or typical weather conditions in a region of the world. The climate in Hawaii is quite different from the climate in Wisconsin. The Hawaiian climate is warm, sunny, and pleasant all year long. The Wisconsin climate is freezing cold in the winter, and hot and humid during the summer.

There are about 12 general climate zones in North America. The two variables that are most important for determining a climate zone are the average temperature throughout the year and the amount of precipitation throughout the year.

This climate map shows the distribution of climate types in North America.

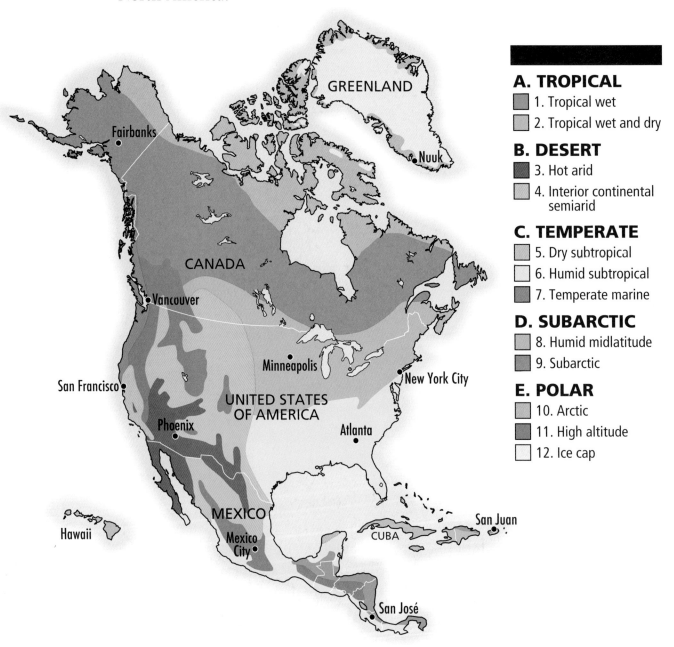

In the Midwest, you can be fairly sure that it will be cold and snowy in January and February and rainy during the summer each year. The same kind of weather will be experienced in Minnesota, Illinois, Connecticut, and Maine. The humid midlatitude climate zone includes the midwestern United States, New England, and the southern part of Canada. This climate zone supports huge diverse forests of deciduous and evergreen trees and all the animals that forests support.

The weather in the southeastern United States is significantly different. Florida, Mississippi, and Louisiana rarely have snow in the winter, and the summers and springs are rainy, hot, and humid. The southern states fall into the humid subtropical climate zone. This zone supports large hardwood forests and many kind of vines.

Humid subtropical zone

The hot arid climate zone in the western United States has predictably warm, dry winters and very hot, dry summers. Arizona and parts of Nevada, Utah, and California are sunny and dry all year. Little rain falls during most of the year. During the summer the temperature can be very high, and thunderstorms can deliver heavy rains that can cause flash floods. The hot arid zone supports a wide diversity of drought-resistant plants, including cactus, mesquite, and yucca, and a host of burrowing and sun-loving animals.

Hot arid zone

Four other climate zones occur in the west (see the climate map). The interior continental semiarid zone is characterized by warm spring and summer weather, cold winters, and summer thunderstorms with the possibility of tornadoes. The semiarid climate supports large expanses of sagebrush and huge grasslands. Land in the interior continental semiarid zone is often used by ranchers to graze livestock.

Interior continental semiarid zone

The high-altitude zone found high in the mountains supports large forests of evergreen trees and provides the right conditions for skiing and other winter sports requiring snow.

Weather in the dry subtropical zone is usually warm and rainy in the winter but hot and dry in the summer. The dry subtropical zone supports oak woodlands, chaparral, and a very diverse community of brush, grasses, and mixed forests. Dry subtropical climates are excellent for farming, fruit orchards, vegetable gardens, and raising livestock.

High-altitude zone

Dry subtropical zone

The temperate marine zone of the Pacific Northwest is cool and wet throughout the year. The climate is strongly influenced by the Pacific Ocean, which keeps the weather cool and moist. This climate zone is characterized by dense forests of large evergreen trees: redwood, fir, pine, and spruce. The moist forests are often home to ferns, mosses, lichens, and fungi. Winters are cool and rainy, while summers are cool and can be foggy and wet.

Two climate zones occur in Alaska. They are the subarctic and the arctic. The climate is extremely cold most of the year, with variable precipitation.

Hawaii has a tropical wet and dry climate, warm and sunny all year long with plenty of tropical rain in many parts of the islands.

Climates vary widely across the country. Many states have only one kind of climate throughout, such as Michigan, Massachusetts, Alabama, and Florida. Other states have two or more kinds of climate. Look at Texas and California. How many climate zones do these states have? So when you are asked what the weather will be like in California, you have to know what part of the state, and what time of year.

Temperate marine zone

Arctic zone

Tropical wet zone

Global Climate Change

Weather is the condition of the atmosphere in a particular place on Earth. Temperature, humidity, and wind describe the weather. The weather is different all over Earth, depending on where you are and the time of year. Climate is the average weather over many years in a region on Earth. The climate in Barrow, Alaska, is very different than the climate in Tahiti. Barrow's climate is extremely cold. The climate on the tropical island of Tahiti is very sunny and warm. Earth's climates are predictable today, but climates have changed many times throughout Earth's history.

Factors Affecting Climate through History

Temperature on Earth is affected by one major energy source, the Sun. The amount of solar energy (heat and light) given off by the Sun is steady day after day, and year after year. But there are many factors that affect the amount of solar energy that transfers to Earth.

An arctic climate

A tropical island climate

76

Ash from a volcano pollutes the air.

One factor that affects the amount of solar energy transferred to Earth is the amount of pollutants in the air. At times in Earth's history, volcanic eruptions, smoke from forest fires, and major impacts by asteroids and comets have put a lot of dust and many gases into the air. These pollutants act like a shield. They block solar energy from reaching Earth's surface. Smoke and dust that block solar energy can cool the climate in large regions of Earth.

The **greenhouse effect** is another factor that affects the amount of solar energy transferred to Earth. Carbon dioxide, methane, nitrous oxide, and water vapor are greenhouse gases. Greenhouse gases in the air act like a mirror. They let through solar energy, which Earth's surface absorbs. Then Earth's surface transfers the energy to the atmosphere. Once the solar energy is in the atmosphere, the greenhouse gases prevent the energy from easily escaping back into space. Heat builds up and the temperature of the atmosphere rises. The trapping of energy in the atmosphere is the greenhouse effect. It impacts climate worldwide.

The Greenhouse Effect

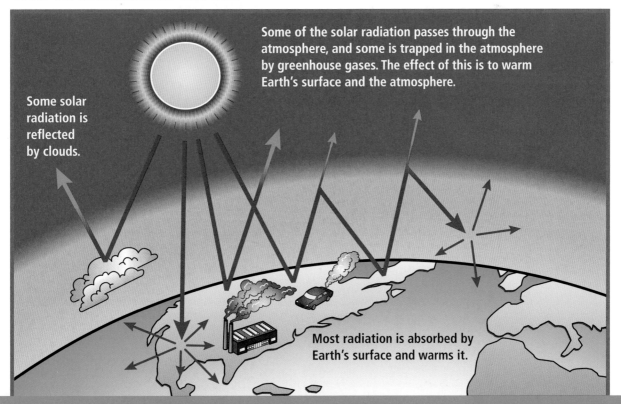

Some solar radiation is reflected by clouds.

Some of the solar radiation passes through the atmosphere, and some is trapped in the atmosphere by greenhouse gases. The effect of this is to warm Earth's surface and the atmosphere.

Most radiation is absorbed by Earth's surface and warms it.

What Is Changing Today?

Today, we are experiencing a period of rapid climate change. The average temperature worldwide has increased about 0.8 degrees Celsius (°C) since 1850. That doesn't sound like much. But think about how much energy it would take to warm all of Earth's atmosphere and the ocean that much. That's a lot of energy.

Scientists have developed climate models that suggest that the global temperature may increase by another 1.5°C–5°C by the year 2100. This temperature change will affect life on Earth and the global climate. Here's how Earth's climate may change.

The Sun rising over Earth

Rising sea level will flood coastal cities.

More farming land will change into desert.

Higher global temperature will cause glaciers and ice sheets to melt worldwide. The arctic polar ice pack will melt completely. Large areas of ice will melt from Greenland and Antarctica. All that melted ice will flow into the ocean and cause the sea level to rise. By 2100, sea level is expected to rise 0.6 meters (m) or more. That is enough to flood many of the low-lying regions of the world. Large parts of Florida, a number of small islands in the Pacific Ocean, and the city of Venice, Italy, would be under water.

As Earth gets warmer, more land will change into desert. Land now used for farming in north central Africa and central Asia will become desert. As a result, the world will produce less food. Food may become more scarce in many parts of the world.

Too much carbon dioxide in the air causes climate change.

What Is Causing Climate Change?

Scientists agree that the major cause of climate change is human activity. The burning of fossil fuels is the number-one human activity affecting climate change. Fossil fuels are the remains of organisms that lived long ago. Over time, those remains changed into oil, coal, and natural gas. When fossil fuels burn, they release carbon dioxide into the air. Humans burn fossil fuels to generate electricity and to power cars.

Carbon dioxide is found naturally in air. In fact, it is essential to life on Earth. Plants use carbon dioxide in the air to produce food by photosynthesis. But since the Industrial Revolution in the 19th century, humans have been releasing more carbon dioxide into the atmosphere than all the plants in the world can absorb. The result is a lot of carbon dioxide in the air. Carbon dioxide is a greenhouse gas. The more carbon dioxide gas we put into the atmosphere, the faster Earth's temperature will rise.

What Can We Do?

So how can we slow climate change? The best way is to stop adding carbon dioxide to the air. But individuals can't make that global decision alone. Governments and big companies need to get involved. For example, the decision to stop burning coal to produce electricity affects many people, because there would be no electricity in their homes. Before we can stop burning fossil fuels to generate electricity, we need to create alternative sources of electricity. Here are some alternative sources for producing electricity.

Wind turbines (windmills) change wind energy into electricity.

Solar energy generates electricity directly with sunlight and solar cells. Solar energy can also produce electricity indirectly by using a large number of curved mirrors as solar collectors. The mirrors focus the Sun's energy on water flowing through a tube. The heated water produces steam that turns electric generators.

A wind farm

A group of solar cells

A system of solar collectors using mirrors

Geothermal energy produces electricity with hot water and steam from volcanic vents to turn generators.

Hydroelectric generators use water flowing through turbines to produce electricity.

Thermonuclear reactions in nuclear power plants create heat to produce steam. The steam turns electric generators to produce electricity.

These alternative sources of electricity are carbon-free. They don't release any carbon into the air. But each alternative source of electricity has its own challenges. If we are going to provide alternative energy to large numbers of homes, it needs to be reliable, safe, and accessible to everyone. We need to develop the technology to convert these primary energy sources (wind, moving water, sunlight, and volcanic vents) into electricity that can be used on a large scale.

A geothermal power plant

A hydroelectric power plant

A nuclear power plant

You can help conserve energy by riding a bike instead of riding in a car.

What Can You Do Right Now?

The most important thing you can do to slow climate change is to conserve energy. Energy used in your home and for transportation is where most of the carbon dioxide in the air comes from. So the next time you need to get from one place to another, don't ask for a ride in the car. Instead, maybe you can walk, or ride your bike or skateboard. Use your own energy instead of carbon energy to get around. And you will get some exercise, too!

At home, you can replace burned-out lightbulbs with compact fluorescent lightbulbs. These use a lot less electricity to produce the same amount of light. If your family needs a new appliance, like a refrigerator, freezer, or water heater, suggest an energy-efficient appliance. You can also adjust the thermostat in your home so it is not too cool or too warm. In the winter, put on a sweater instead of turning up the heat. In the summer, wear lightweight clothes, and enjoy the warmth.

What other ways can you conserve energy? Become an energy detective at school. Find ways your school can conserve energy.

A compact fluorescent lightbulb

Science Safety Rules

1. Listen carefully to your teacher's instructions. Follow all directions. Ask questions if you don't know what to do.
2. Tell your teacher if you have any allergies.
3. Never put any materials in your mouth. Do not taste anything unless your teacher tells you to do so.
4. Never smell any unknown material. If your teacher tells you to smell something, wave your hand over the material to bring the smell toward your nose.
5. Do not touch your face, mouth, ears, eyes, or nose while working with chemicals, plants, or animals.
6. Always protect your eyes. Wear safety goggles when necessary. Tell your teacher if you wear contact lenses.
7. Always wash your hands with soap and warm water after handling chemicals, plants, or animals.
8. Never mix any chemicals unless your teacher tells you to do so.
9. Report all spills, accidents, and injuries to your teacher.
10. Treat animals with respect, caution, and consideration.
11. Clean up your work space after each investigation.
12. Act responsibly during all science activities.

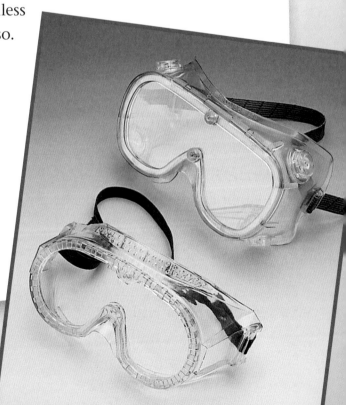

Glossary

absorb to soak in

air the mixture of gases surrounding Earth

air pressure the force exerted on a surface by the mass of the air above it

anemometer a weather instrument that measures wind speed with wind-catching cups

atmosphere the layer of gases surrounding Earth. The layers include the troposphere, stratosphere, mesosphere, thermosphere, and exosphere.

barometer a weather instrument that measures air pressure

blizzard a severe storm with low temperatures, strong winds, and large quantities of snow

climate the average or typical weather conditions in a region of the world

cloud tiny droplets of water, usually high in the air

cold front the contact zone where a cold air mass overtakes a mass of warm, moist air

condensation the process by which water vapor changes into liquid water, usually on a surface

conduction the transfer of energy from one place to another by contact

conserve to use carefully and protect

convection current a circular movement of fluid (such as air) that is the result of uneven heating of the fluid

dew water that condenses on a surface when the temperature drops at night

drought a less-than-normal amount of rain or snow over a period of time

Earth the third planet from the Sun, known as the water planet

energy transfer the movement of energy from one place to another

evaporation the process by which liquid water changes into water vapor

exosphere the layer of the atmosphere above the thermosphere. The exosphere makes the transition from the atmosphere to space.

fog water droplets that condense from the air close to the ground

forecasting predicting future events or conditions, such as weather

fossil fuel the preserved remains of organisms that lived long ago and changed into oil, coal, and natural gas

front the leading edge of a moving air mass

frost frozen condensation

gas a state of matter with no definite shape or volume; usually invisible

greenhouse effect when heat builds up in the atmosphere and causes Earth's temperature to rise

hail precipitation in the form of small balls or pellets of ice

humidity water vapor in the air

hurricane a severe tropical storm that produces high winds

hygrometer a weather instrument that measures humidity

kinetic energy energy of motion

liquid a state of matter with no definite shape but a definite volume

mesosphere the layer of the atmosphere above the stratosphere

meteorologist a scientist who studies the weather

ozone a form of oxygen that forms a thin layer in the stratosphere

photosynthesis a process used by plants and algae to make sugar (food) out of light, carbon dioxide, and water

precipitation rain, snow, sleet, or hail that falls to the ground

predict to estimate a future event based on data or experience

radiant energy energy that travels through air and space

radiation energy that travels through air and space

radiosonde an instrument sent into Earth's atmosphere to measure temperature, pressure, and humidity

rain liquid water that is condensed from water vapor in the atmosphere and falls to Earth in drops

recycle to use again

reflect to bounce off an object or surface

renewable resource a natural resource that can replenish itself naturally over time. Air, plants, water, and animals are renewable resources.

season a time of year that brings predictable weather conditions to a region on Earth

severe weather out-of-the-ordinary and extreme weather conditions

sleet precipitation in the form of ice pellets created when rain freezes as it falls to Earth from the atmosphere

snow precipitation in the form of ice crystals grouped together as snowflakes

solar energy heat and light from the Sun

stationary front the contact zone where a warm air mass and a cold air mass come together and stop moving

stratosphere the layer of the atmosphere above the troposphere. The ozone layer is in the stratosphere.

Sun the star at the center of the solar system

temperature a measure of how hot the air is

thermometer a weather instrument that measures temperature

thermosphere the layer of the atmosphere above the mesosphere

thunderstorm severe weather that results from cold air flowing under a warm, humid air mass over the land

tornado a rapidly rotating column of air that extends from a thunderstorm to the ground. Wind speeds can reach 400 kilometers (km) per hour or more in a tornado.

transpiration the process by which water escapes into the air through plants

troposphere the layer of the atmosphere that begins at Earth's surface and extends upward for 9 to 20 km. Weather happens in the troposphere.

warm front the contact zone where a warm air mass overtakes a cold air mass

water cycle the global water-recycling system. Water evaporates from Earth's surface, goes into the atmosphere, and condenses. It returns to Earth's surface as precipitation in a new location.

water vapor the gaseous state of water

weather the condition of the air around us. Heat, moisture, and movement are three important variables that describe weather.

weather map a map that shows weather data in symbols

weather variable data that meteorologists measure. These include temperature, wind speed and direction, air pressure, cloud cover, and precipitation.

wind air in motion

wind meter a weather instrument that measures wind speed with a small ball in a tube

wind vane a weather instrument that measures wind direction

Index

A
Absorb, 6, 10–11, 17–19, 26, 85
Air, 3–6, 14–16, 21–26, 27–31, 50, 66, 85
Air pressure, 9, 13, 14, 15, 54, 57, 62, 63, 64, 68, 70, 85
Anemometer, 15, 16, 85
Atmosphere, 3–13, 49, 85
Atmospheric measurements, 64, 66

B
Barometer, 15, 85
Blizzard, 58, 85

C
Carbon dioxide, 3–6, 26, 77, 80
Climate, 60, 71–75, 85
Climate change, 60, 76–83
Cloud, 5, 44, 50, 51, 54, 61, 68, 85
Cold front, 66–68, 85
Compass, 16
Condensation, 5, 15, 43–46, 55, 85
Conduction, 21–26, 85
Conserve, 40, 83, 85
Convection, 27–31, 56
Convection current, 29, 30, 31, 54, 85

D
Day/night, 30
Density, 15, 28, 29
Dew, 45, 50, 51, 85
Drought, 53, 59–60, 85

E
Earth, 85
    atmosphere, 3–13, 49
    climate, 60, 71–83, 85
    heating of, 17–20, 21–26, 28, 29, 31
Energy. See also Heat; Kinetic energy; Solar energy
    role in weather changes, 54
    objects in motion, 22
    photosynthesis, 6
    uneven heating, 17
    wind, 31, 32–33
Energy transfer, 23–26, 28, 85
Equator, 9, 66
Evaporation, 5, 43, 49, 50, 54, 85
Exosphere, 8, 13, 85

F
Fog, 44, 45, 50, 85
Forecasting, 62, 68–70, 85
Fossil fuel, 33, 80, 85
Front, 66–68, 85
Frost, 46, 85

G
Gas, 3–13, 15, 22, 43, 50, 85
Geothermal energy, 82
Greenhouse effect, 77, 85

H
Hail, 51, 86
Heat, 6, 10, 12, 17–20, 21–26, 28, 29, 31, 43, 54
Humidity, 9, 14, 15, 16, 54, 62, 64, 86
Hurricane, 31, 53, 55, 86
Hydroelectric energy, 82
Hygrometer, 15, 86

K
Kinetic energy, 22, 24, 25, 43, 86

L
Liquid, 14, 22, 43, 50, 52, 86

M
Matter, 17, 22
Mesosphere, 8, 12, 86
Meteorologist, 14, 62, 63–66, 70, 86

N
National Weather Service, 70
Nuclear energy, 82

O
Ocean, 50, 51, 58, 61
Oxygen, 4, 6, 26
Ozone, 4, 6, 11, 86

P
Pattern, predictable, 71
Photosynthesis, 6, 86
Precipitation, 5, 51, 54, 62, 63, 71, 86
Predict, 62, 86

R
Radiant energy, 24, 26, 86
Radiation, 21–26, 86
Radiosonde, 64, 86
Rain, 5, 15, 49, 51, 52, 53, 54, 59, 67, 68, 86
Recycle, 51, 60, 86
Reflect, 17, 19, 86
Renewable resource, 42, 86

S
Safety rules, 84
Season, 9, 86
Severe weather, 53–61, 86
Sleet, 51, 86
Snow, 5, 49, 51, 58, 86
Solar electricity, 38–39
Solar energy, 17, 24, 86
    greenhouse effect, 76–77, 81
    hot and cold weather, 58
    technology, 34–42
    temperature change, 12
    uneven heating, 18, 19, 20
    wind and convection, 28
Stationary front, 67, 86
Stratosphere, 8, 11, 86

Sun, 86
    atmosphere, 3
    heat from, 10–12, 66
    heat transfer, 22–26
    impact on Earth's weather, 17, 54, 76
    role in water cycle, 50
Sunlight, 6, 17–20, 81

T
Telkes, Maria, 40
Temperature, 86
    climate zone, 71
    differences in, 8, 9–13, 19
    extreme, 7
    global climate change, 76–79
    hot and cold, 58
    measuring, 62, 63, 64, 66
    role in water cycle, 51
    weather instruments, 16
    weather variable, 14, 54, 71
Thermometer, 14, 54, 86
Thermosphere, 8, 12, 87
Thunderstorm, 31, 53, 56, 87
Tornado, 31, 53, 57, 66, 87
Transpiration, 51, 87
Troposphere, 8, 9–10, 87

U
Ultraviolet radiation, 6, 11

W
Warm front, 67–68, 87
Water
    condensation, 43–46
    location of, 47
    recycling, 60
    temperature of, 19, 20, 28, 55
Water cycle, 48–52, 87
Water vapor, 3–5, 15, 43–46, 50, 51–52 55, 61, 77, 87
Weather, 14–16, 87
    changes in, 54, 66
    extremes, 58–60
    forecasts, 62, 68–70
    hot and cold, 58
    role of ocean in, 61
    severe, 53–61
    troposphere and —, 9
Weather balloon, 9, 64
Weather map, 62–70, 87
Weather variable, 9, 14, 54, 63, 71, 87
Wind, 87
    convection and —, 27–31
    energy, 32–33
    forecasting, 68–70
    role in water cycle, 51
    severe weather, 53, 55–57
    speed/direction, 15, 16, 30, 53, 57, 62, 63, 64
    weather variable, 9, 14, 54
Wind meter, 15, 87
Wind vane, 16, 87